This is a Parragon Book.



Parragon
13 Whiteladies Road, Clifton,
Bristol BS8 1PB

Produced by The Templar Company plc,
Pippbrook Mill, London Road, Dorking,
Surrey RH4 1JE

Written by Robert Snedden
Illustrated by Peter Bull Art Studio
Series Designer Mark Summersby



Printed and bound in China

ISBN 0 7525 1692 2

FACTFINDERS

# GARDEN BIRDS

PARRAGON

## Titles in this series include:

Astronomy
Card Games
Clans & Tartans
Fastest Cars
Flags of the World
Freshwater Fish
Garden Birds
Handguns & Small Arms
Historic Britain
Human Body
Insects
Inventions
Modern Combat Aircraft
Natural Disasters
Pond Life
Rocks & Fossils
Wild Flowers
World War II Combat Aircraft

# Contents


Introduction . . . . . . . 8
Thrushes . . . . . . . . . 12
Starling . . . . . . . . . . 18
Warblers . . . . . . . . . 19
Crests . . . . . . . . . . . 24
Flycatcher . . . . . . . . 26
Hoopoe . . . . . . . . . . 27
Tits . . . . . . . . . . . . . 28
Waxwing . . . . . . . . . 32
Wren . . . . . . . . . . . . 33
Crows . . . . . . . . . . . 34
Sparrow . . . . . . . . . . 38
Dunnock . . . . . . . . . 39
Nuthatch . . . . . . . . . 40
Treecreeper . . . . . . 41
Finches . . . . . . . . . . 42
Bunting . . . . . . . . . . 47
Pigeon . . . . . . . . . . . 48
Cuckoo . . . . . . . . . . 49
Woodpeckers . . . . . 50
Swift . . . . . . . . . . . . 52
Martin . . . . . . . . . . . 52
Swallow . . . . . . . . . . 53
Owls . . . . . . . . . . . . 54
Raptors . . . . . . . . . . 56
Kingfisher . . . . . . . . 58
Wagtails . . . . . . . . . 59
Heron . . . . . . . . . . . 60
Duck . . . . . . . . . . . . 61
Gulls . . . . . . . . . . . . 62
Less Common Birds . . . . . . . . . . . . 66
Encouraging Garden Birds . . . . . . . . . . 80
Bird Tables & Feeding . . . . . . . . . 82
Nest Boxes in the Garden . . . . . . . . . . 85
Garden Design & Plants . . . . . . . . . . . 87
Glossary . . . . . . . . . 89
Index . . . . . . . . . . . . 93

# Introduction

Birds are among the most common visitors to our gardens and so it is no surprise to consider that almost everyone with access to a garden must sometimes wonder about the birds they see there, or perhaps hear as part of the glorious dawn chorus.

There are many other animal visitors to gardens—everything from foxes and the occasional deer (in rural areas) to mice, moles and the neighbour's perhaps unwelcome cat—but it is often true that birds are the most noticeable and, often, one of the most numerous. They are also some of the most interesting, since regular bird-watching can reward you with an insight into many more facets of their lives than is possible with other animal species. In a garden you can watch birds feed, court, make nests, raise their young and also watch out for the progress of the offspring themselves. It is also true that in most areas it is fairly easy to encourage birds into your garden and we've included a special section at the back of the book (see pages 80 to 88) to help you do just that.

But once you've started to watch these feathered visitors to your home, how do you tell which species is which? Some birds will be familiar to almost everyone—the ubiquitous sparrows and starlings, the friendly robin, the distinctive thrush. But what of the others?

The way a bird looks, its song and its habits will all help you to identify it. Look for field marks—the important visual features that distinguish the bird from others. A good pair of binoculars is invaluable for this, as not many birds are going to let you walk right up to them. Make sure you choose ones that are comfortable and light enough to hold. Buy the best lenses you can afford – good binoculars will reward the money spent on them, bad ones may just frustrate you. A magnification of x8 is about right for most bird-watching activities.

Remember to show consideration for the birds you are watching. Do not, for example, get in the way of a bird that is trying to bring food to its chicks or material to its nest. If a bird seems to be agitated by your presence then leave it alone. Instead, watch from a reasonable distance, or from the comparative comfort of your kitchen window.

If you want to do something to help safeguard birds and their environment, or are interested in finding out more about bird-watching in general, then you might consider joining the Royal Society for the Protection of Birds (RSPB), or its junior section, the Young Ornithologists' Club. The address for both is:

The Lodge
Sandy
Bedfordshire
SG19 2DL

# THRUSHES

**Blackbird** A smallish (25cm) bird that is very common in gardens at any time of year, the blackbird is a member of the thrush family. The glossy black male with his bright yellow bill is unmistakable, while the female bird is brown above, lighter brown below with dark mottling. The blackbird's song is mellow and fluting, although it has a variety of other calls. The blackbird is a year-round resident in Britain, although some northern European blackbirds winter in the Mediterranean.

**Song Thrush** The commonest European thrush, the song thrush is smaller (23cm) than the similar mistle thrush, with warmer brown upper parts and smaller, narrower spots. Its loud, clear song is made up of a variety of repeated phrases, generally delivered from a perch high in a tree or on a building. It is a migratory bird, which spends the winter in north Africa or southern Europe, returning in March.

**Mistle Thrush** Widespread throughout the British Isles, but less common than the smaller song thrush (above), the mistle thrush (26.5cm) can be identified by its larger breast spots. Also, a flash of white can be seen under its wings when it is in flight. Its ringing song, with a number of short phrases, lacks the song thrush's repetitiveness. It is a migratory bird, and can arrive at its breeding grounds as early as February. It can be most often found in woodland, parks and large gardens.

**Redwing** A small (21cm), short-tailed thrush, the redwing is similar in appearance to the song thrush, but can be identified by its reddish-brown flanks and creamy-white eyestripe. The bird's call note is a thin 'see-ip'. Generally, the redwing is a winter visitor to Britain, when it is fairly common. It arrives here in September and leaves in March or April, although there are some pairs of birds which breed in the forests of birch and conifers in the north of Scotland. It can often be seen in flocks, stripping hedgerows and garden plants of berries in the autumn, usually in the company of blackbirds and other thrushes. Growing berry-producing plants, such as rowan, hawthorn and yew, can help encourage such birds to become regular visitors to your garden.

**Robin** Perhaps one of the best-known of all British birds, the tiny (13cm) robin, with its distinctive orange-red breast, is a notably tame and common visitor to town and city gardens. Its song is varied and tuneful with short warbling phrases. Robins are highly territorial and will not usually tolerate other robins living too close by. They have even been known to attack other small red objects in their desire to protect their "patch" from an intruder. Many live in Britain all year round, although there are others who come here in the autumn from northern and eastern Europe to spend the winter. Robins feed chiefly on insects, berries and worms and will often appear if you are digging in the garden in the hope of picking up a tasty morsel.

**Redstart** An occasional visitor to suburban gardens, the little (14cm) redstart usually frequents woods, parkland, orchards or places where there are scattered trees. Look for it also perched on the roofs of sheds, houses and other buildings. An agile bird, it can be distinguished by its chestnut tail and orange breast and flanks, and by the up and down flickering movement of its tail. Its song is a brief melodious warble. A summer visitor to Britain from Africa, the redstart arrives between April and May and leaves here any time from July to October. It has been known to use nest boxes but more usually builds a nest of plant material in a wall crevice, among the rafters of a roof or in an old tree trunk.

**Nightingale** A small (17cm) thrush, the nightingale has a rich, varied and vigorous song that can often be the best indication of its presence. It is most often heard before dawn or at dusk when most other birds are silent and its magnificent voice can carry for up to a mile. Looking something like a large brown robin in appearance, the nightingale is a highly secretive species, tending to skulk under cover of foliage. It is often found where there are trees and plenty of undergrowth, often singing from the safety of dense cover. A migratory bird, the nightingale is a summer visitor to southern England, wintering in tropical Africa. It is not found in Scotland.

# STARLING

**Starling** The starling is another small (21cm) bird that can commonly be found in parks and gardens in towns and suburbs. The adult starling has an iridescent black plumage that becomes spangled with pale spots in winter, when its beak turns from yellow to black. The starling's song is a rambling melody with some elements imitating the calls of other birds, while its call is a harsh 'tcheer'. It is a year round resident in Britain.

# WARBLERS

Garden Warbler

**Garden Warbler** Warblers, like the small (15cm) garden warbler, are insect-eating land birds. Their calls are often the best way of distinguishing one species from another and they are often found where there is plant cover of some sort. The garden warbler's *lack* of distinct features in its uniform plumage aids identification. Its song is a sweet, sustained musical warble and may be imitated by the blackcap. A migratory bird, the garden warbler is a summer visitor to Britain, where it is seen in woods, parks and large gardens.

**Whitethroat** A small (14cm), active warbler, the whitethroat is most usually seen among hedgerows and scrub. It is particularly fond of brambles and nettles so leaving a wild, uncultivated patch of land in your garden may help to encourage its presence. The male bird can be identified by its grey head and white throat in the breeding season; reddish-brown edges to the wings are a distinctive feature in both sexes. The whitethroat's song is a distinctive, staccato, rapidly delivered warble. A summer visitor to Britain, the whitethroat spends the winter in Africa.

**Blackcap** A small (14cm) warbler, the blackcap is a lover of forests, overgrown gardens and parks and is easily identified by the dark colour of its crown which is black in the male bird and brown in the female. Its song is generally higher pitched and with shorter phrases than that of the garden warbler, which it may mimic. British blackcaps usually spend the spring and summer here, migrating southwards in August or September to spend the winter in Africa or southern Europe. Adult birds feed chiefly on insects and their larvae, although they are also fond of berries, especially raspberries and elderberries.

Male Blackcap

Female Blackcap

**Chiffchaff** The little (11cm) chiffchaff is very similar to the willow warbler in appearance, but its song is quite distinctive, consisting of the repeated monotonous 'chiff-chaff' that gives rise to its name. Its call is a simple 'hweet', similar to that of the willow warbler. The chiffchaff is principally a summer visitor to the Britain Isles, usually arriving around April and leaving any time from August to mid-October. The chiffchaff is seen in similar habitats to the willow warbler, but prefers an environment with fewer bushes and taller trees.

**Willow Warbler** The small (11cm) willow warbler is the commonest northern European warbler, and is very difficult to distinguish from the chiffchaff (opposite) other than by its song. The willow warbler's song consists of a series of descending notes, gaining in loudness before fading away again. The 'hooeet' call is more definitely two notes than that of the chiffchaff. A summer visitor to Britain, arriving in April and leaving around September, it is generally found in areas with plenty of trees and bushes.

# Crests

**Goldcrest** The tiny (9cm) goldcrest shares with the firecrest the honour of being Britain's smallest bird. The goldcrest can be identified by its dull green upper parts and also by its black-bordered orange-yellow crest. The goldcrest's call is a shrill 'zeec-zeec-zeec' while its song is a thin and very high-pitched 'cedar-cedar-cedar', ending with a flourish. It is most usually found in woods and large gardens, especially around conifers. A year-round resident, the goldcrest feeds mainly on insects.

**Firecrest** This tiny bird (9cm) can be distinguished from the equally small goldcrest by its whiter underparts, black-and-white eyestripes and by a bronzy patch on the side of its neck. The firecrest's song is lower and less thin than the goldcrest's, being basically a series of repetitions of the 'zit, zit, zit' call. In Britain, the firecrest is confined to the south of England, where it is found in a similar habitat to that of the goldcrest, although more often in bushes and lower cover.

# FLYCATCHER

**Spotted Flycatcher** A small (14cm) bird, the spotted flycatcher can often be seen sitting upright on some high, exposed perch, from which it makes frequent forays after insects in a fluttering, erratic flight. Usually found near forest edges and in parks, orchards and gardens, the flycatcher particularly likes to use nest boxes. Seldom seen on the ground, the flycatcher's call is a shrill 'tzee'. A summer visitor, the flycatcher usually visits these shores from mid-April to October.

# Hoopoe

**Hoopoe** The medium-sized (28cm) hoopoe is easy to identify as its long curved bill, pinkish-brown plumage, barred wings and prominent crest are unmistakable. This bird is a rare summer visitor to the south of England where it is found in farmland, vineyards, parks and gardens. You are more likely to hear its distinctive 'oop-oop' call than you are to see it.

**Coal Tit** The coal tit is a tiny (11.5cm) bird identified by a combination of its black cap and conspicuous white nape patch. The white patches are tinged yellow in the young bird. The coal tit's song is a piping 'teechu, teechu, teechu'. Although it is most common in coniferous woods, it can also be found in woods of all kinds, in parks and in gardens. Birds from northern Britain tend to migrate south for the winter.

**Long-tailed Tit** The long-tailed tit is a small (16.5cm) bird and, like all tits, feeds on insects. Often seen in groups of several birds in gardens, it is less common than some of the other tits listed here. The bird's pink, black and dull white plumage and its very long, narrow tail help easy identification. Its calls are chiefly a soft 'tupp' and a penetrating 'tsirrup'. The long-tailed tit's flight is somewhat laboured and undulating. It is rarely seen on the ground. It is frequently found in broad-leaved woodland and scrub and lives in Britain all year round.

**Blue Tit** The bright blue and yellow plumage of the tiny blue tit (12cm) is its most distinctive feature, though male blue tits are normally brighter than the females. The bird's song is a cheerful liquid trill and there is also a typical scolding 'tsee-tsee-tsee, tsit'. It is very common in town parks, gardens, orchards and broad-leaved woods, where it appears both in groups and alone. A year-round resident of Britain, it is one of the most frequent visitors to bird tables in the winter and is very partial to 'bird pudding'. It will also happily nest in nest boxes if they are provided, as well as tree cavities.

**Great Tit** A small (14cm) bird, the great tit is the largest of the tits. It can readily be identified by its black-and-white head and by the broad black band which runs down its bright yellow breast. The bird's chief song is a metallic 'teechu-teechu-teechu'. The great tit is common in all kinds of woodlands, orchards, parks and gardens and, like its relative the blue tit, will often take up residence in nest boxes as well as being a frequent visitor to bird tables. It feeds its young mainly on caterpillars–watch for it in the spring when parent birds might make as many as 500 trips a day to and from its nest ferrying food to their chicks.

# Waxwing

**Waxwing** A small (18cm) bird, no other bird has the same combination of brown colouring, prominent crest, yellow and red wing markings and yellow-tipped tail as the waxwing. Its grey rump is also a distinguishing feature in flight. The song is a weak 'tsirrrr'. Living mainly in the far north, the waxwing is a winter visitor to Britain, where it is almost always seen near the berried shrubs in parks and gardens that provide its main food source.

# Wren

**Wren** The tiny (9.5cm) wren is the smallest brown bird of the region with a distinctive, cocked-up little tail. It is a highly active bird, constantly on the move as it hunts for the insects, grubs, spiders and small seeds that make up its diet. It flies short distances with a rapid whirring of wings, but may be more often seen hopping through the undergrowth more like a mouse or other small mammal. Its clear, warbling song is remarkably loud for a bird of such tiny size and it is found across a broad range of habitats, from mountain tops to suburban gardens. However, it is most commonly seen where there is woodland and dense undergrowth. It is a year-round resident of Britain.

# Crows

**Jay** The medium-sized (34cm) jay is a colourful crow with pinkish-brown plumage, black-and-white crown and a distinctive wing patch striped in bright blue, black and white. The bird's crown can be erected to form a crest. Its most distinctive call is a scolding 'skaak, skaak', and it is often found where there are scattered trees, especially of oak. In Britain it is a year-round resident and, after the birds have reared their young in April or May, they gather together in flocks made up of several families.

**Magpie** A medium-sized (46cm) member of the crow family, the magpie has glossy, pied plumage and a long, diamond-shaped tail that are particularly distinctive. It is well known for its habit of picking up shiny objects and then hiding them, and is also notorious for feeding on the eggs and young of other birds—a fact that led to its persecution, particularly by gamekeepers, throughout the 19th century. The magpie's flight is slow with rapid wing beats and its usual call is a harsh chattering or chuckling note. It is most often found among scattered trees, but it can often be seen around the edges of garden ponds or perching in tall trees along the garden boundary. It commonly breeds in towns and cities as well as in its more usual countryside habitat.

**Jackdaw** The jackdaw is the smallest (33cm) black crow, with a distinctive grey nape. It walks with a jaunty action and its flight is fast with rapid wingbeats. Its most usual call is a high pitched 'chack', but it will also imitate the calls of other birds. Feeding on seeds, jackdaws will often come into gardens where there is fruit such as cherries or strawberries.

**Carrion Crow** The crow is a medium-sized (47cm) bird. There are two subspecies: the all-black carrion crow and the hooded crow, which is a grey bird, with black head, wings and tail. Distinguished from the rook by the fully-feathered base of its bill and more compact plumage, the crow feeds mainly on the ground, moving with ungainly hops. Its flight is slow with regular, deliberate wingbeats and its usual call is a croaking 'kraah'. It is found in all types of country, farmland, parks and suburban gardens.

Carrion Crow

Hooded Crow

**Rook** The rook is a medium-sized (46cm) crow. The adult rook is distinguished from the carrion crow by its bare face patch and the baggy appearance of its thighs, and it is distinguished from the jackdaw by its greater size. Its usual call is a harsh 'kaah' and it is usually found in flocks on farmland and grassland where there are scattered trees, roosting in treetops overnight. It also nests communally, many birds returning to their tree-top rookeries in early spring to rebuild their nests. One of the largest rookeries ever found in Britain contained over 2,000 individual nests.

# Sparrow

**House Sparrow** The small (14.5cm) house sparrow is a very common bird in cities and gardens throughout Europe and Asia and has been successfully introduced to almost all other parts of the world. Sparrows are often colonial nesters, making their round, untidy nests in the branches of trees or behind beams and drainpipes, but they are quite happy to make use of nest boxes as well. The male has a dark grey crown, a black throat and greyish-white cheeks while the female is uniformly brown with a single pale wingbar. Mainly a ground feeder, the house sparrow is a noisy, gregarious bird with a broad range of cheeps and chirps and it is a year-round resident in Britain.

# DUNNOCK

**Dunnock** The small (14.5cm) dunnock is an inconspicuous bird, which is also known as the hedge sparrow. Similar in appearance to a female house sparrow, the dunnock is distinguished by its grey head and underparts and by its thin bill. Its call is a thin 'tseep', its song a high-pitched but subdued warbling. It is very much a ground feeder and is found in all sorts of bushy places and in broad-leaved and coniferous woodland. It is one of the commonest garden birds.

# NUTHATCH

**Nuthatch** The small (14cm) nuthatch is a distinctive bird, with a black eyestripe, blue-grey upper parts and buff underparts with chestnut flanks. It moves up and down tree trunks in jerky leaps. Unlike the treecreeper, it does not use its tail to aid it when it is climbing. Its chief call is a ringing 'chwit-chwit'. Nuthatches are always found in trees, in forests, woods, parkland or large gardens and, as year-round residents, they can often be seen visiting bird tables in the wintertime.

# TREECREEPER

**Treecreeper** The little (12.5cm) treecreeper is the only British perching bird with a long downward-curving bill. It most often inhabits areas where there are coniferous trees and it is rarely seen away from tree trunks where it will creep up and down in search of insects. Its stiff, pointed tail feathers are pressed against the trunk to help with climbing. Its song is a thin and high-pitched 'tee-tee-tee-titidooee'. It is a year-round resident.

# FINCHES

**Chaffinch** The small (15cm) chaffinch is the commonest of the European finches. The male is unmistakable, with its slate-blue head, chestnut mantle and pinkish-brown underparts, while the female is a pale, yellowish-brown. A broad, white shoulder patch is prominent in both sexes in flight. The chaffinch's song is a cheerful, rattling affair and these birds can often be seen as well as heard in woods and forests, farmland, orchards and parks. They are frequent visitors to gardens everywhere and will readily feed from bird tables.

**Bullfinch** The bullfinch is a small (14.5-16cm) bird. The male is quite unmistakable, a stout bird with grey back, black cap and rose-red underparts, the female being somewhat duller and pinker. The usual call is a penetrating low whistled 'deu', which is often the only indication of this shy bird's presence. The bullfinch has no song as such and is found in woodland, scrub, orchards and large gardens where it will devour the buds of flowering trees, especially fruit trees, in the spring.

**Greenfinch** The little (14.5cm) greenfinch is a common bird that can be identified by its yellow-green coloration, its stout bill and also by the bright yellow patches on its wing primaries. The female is slightly duller than the male. The greenfinch's usual call is a rapid, twittering 'chichichichit' and it is found in areas with scattered trees, such as orchards, gardens and parks. In the wintertime, greenfinches are regular visitors to bird tables, particularly if peanuts are on the menu. They can be quite aggressive, driving away other species from nut dispensers and feeding trays. Greenfinches use their strong bills to break open the seeds of various plants, trees and shrubs on which they feed year-round .

**Goldfinch** The adult goldfinch is a distinctive, small (12cm) bird, with a brilliant yellow band on its predominantly black wings and a red, white and black head. Its dancing flight is reminiscent of a butterfly. The goldfinch's song is a liquid twittering with variations of the 'tswitt-witt' call. Goldfinches are found especially in orchards, but will also nest in parks and gardens, although they are shy birds. It is often more likely that you will hear this bird rather than actually see it. Watch for goldfinches particularly wherever there are seed-producing plants—they are experts at extracting seeds from thistle heads, one of their favourite sources of food.

**Redpoll** The tiny (13-15cm) redpoll, another member of the finch family, can be distinguished by the adults black chin, crimson forehead and pink breast. The bird's 'chuch-uch-uch' flight note also aids identification. Its song, a trill based on the flight note, is often delivered during a circling song flight. Redpolls are found in coniferous and birch woods, scrub and also in large gardens where, like most finches, they feed on seeds. In the winter they will often visit parks and gardens in groups, particularly those with birch or other seed-bearing trees.

**Hawfinch** The small (18cm) hawfinch is large for a finch. It has an outsize bill, grey-blue in summer, yellow in winter, which makes it unmistakable, as do the bird's thick neck and short tail. A secretive bird, its note is a clipped 'tzik' or 'tzik-tzik'. It rarely sings. It is most usually found in broad-leaved and mixed woods, orchards and large gardens.

# BUNTING

**Yellowhammer** The small (16.5cm) yellowhammer is a fairly common bunting that often visits suburban gardens during the winter. The male is the only mainly yellow bird with chestnut back and rump in Britain. The female is less yellow and can be identified by her chestnut rump. White outer tail feathers can be seen in flight in both sexes. The yellowhammer's high-pitched song is traditionally rendered 'a little bit of bread and no cheese'.

# PIGEON

**Woodpigeon** The woodpigeon is a medium-sized (41cm), heavily-built bird, which can readily be identified by the prominent white patches on its wings and neck, and its habit of flying off with a noisy clatter of wings if disturbed. Its call is a repeated 'coo-coo-coo coo-coo'. It is found in wooded country, parks and gardens where there is dense undergrowth and, in Britain, is usually a migrant bird, appearing around the middle of March.

# Cuckoo

**Cuckoo** The medium-sized (33cm) cuckoo is similar in appearance to a bird of prey and superficially like a sparrowhawk, but can be distinguished by its pointed wings and tail marked and tipped with white. It is found in a wide range of habitats, from moors and heaths to town parks, and is famous for laying its eggs in the nests of other birds. The call of the male is the well-known 'cooc-coo'. It is a summer visitor to Britain, arriving around April and leaving in July and August.

# Woodpeckers

Woodpeckers are always found near trees, which they climb over in search of insects under bark and in rotten wood. Their stiff tails give support on vertical surfaces and they have two toes pointing forwards and two backwards, to give better grip. Most species 'drum' by tapping rapidly with the bill on dead branches.

**Green Woodpecker** This medium-sized (32cm) bird can be distinguished by its green plumage, red crown and a yellowish rump, seen in flight. It is the largest of the British woodpeckers. In spring, its call is a loud 'queu-queu-queu'. The green woodpecker does not drum as often as other woodpeckers. These birds can most often be found in broad-leaved and mixed forests, and also in parks and large gardens. In some places it is known as the yaffle, because of its laughing call.

**Great Spotted Woodpecker** The presence of this small (23cm), handsome black-and-white woodpecker is often announced by its frequent drumming, with eight to ten blows in the space of a second. It is much larger than the lesser spotted and much smaller than the green woodpecker. It is also distinguished from the lesser spotted by its red under-tail coverts and nape, and its white shoulders. It has a distinctive 'bounding' flight. The main call is a 'tchick, tchick'. It is found in woods, parks, orchards and large gardens.

**Lesser Spotted Woodpecker** This is the smallest (14.5cm) of the region's woodpeckers, being no bigger than a sparrow, with a distinctive barred pattern on its upper parts that readily distinguishes it from the larger great spotted, as does its red cap. The drumming is not so loud as that of the great spotted but is longer lasting, with between 10 and 30 blows in two seconds. It tends to flutter from branch to branch in the trees and its usual call is a shrill, unvarying 'pee-pee-pee pee-pee'. It is found in mixed forests, parks and large gardens in England and Wales.

# SWIFT

**Swifts** Small (16.5cm) birds, swifts spend most of their lives in the air, almost never landing on the ground. They have narrow, scythe-like wings, feed on flying insects and can be distinguished from swallows by their short tails, long curved wings, and uniformly sooty-brown plumage. They are often seen in noisy parties, squealing loudly as they chase insects high over towns. Their nests are commonly built beneath the eaves of tall buildings.

# MARTIN

**House Martin** The small (12.5cm) house martin is, as its name suggests, very common in towns and villages as it frequently makes its nest under the eaves of houses. Its white rump, contrasting strongly with blue-black back and wings is the main identifying feature. House martins are summer visitors to Britain.

# Swallow

**Swallow** Swallows are small (19cm) birds which can be identified by their long tail-streamers and chestnut throats and foreheads: their song is a pleasing twittering warble. They are summer visitors to Britain and, when their young have fledged, they form large flocks which roost amongst the reeds in ponds, readying themselves for the long flight to their winter grounds in tropical or southern Africa. Swallows build cup-shaped nests of mud cemented together with saliva, often inside barns and other farm bui.ldings.

# Owls

Owls are primarily nocturnal birds of prey, although some may be seen hunting by day. They have short tails, long, rounded wings and feathered legs and feet. The head is large with a flattened face, eyes facing forward and a short, hooked bill.

**Barn Owl** This medium-sized (34cm) owl (left) has distinctive orange-buff upper parts and a pure white face and under parts. It often makes its nest in old buildings and hunts rodents over open country. Its call is a blood-curdling screech.

**Little Owl** The small (22cm) little owl is often seen during the day, and its grey-brown plumage is distinctive. Low in flight, with a bounding action, it is found in open country and woodland and its call is a mewing 'kiew, kiew'

**Tawny Owl** The 'hoo-hoo-hoo hooo' cry of the medium-sized (38cm) tawny owl is one of the most familiar of bird calls. Distinguished from the barn owl by its dark face and under parts, and from the smaller long-eared owl by its lack of ear tufts, the tawny owl is rarely seen during the day. It lives in open woodland and parkland, but can sometimes be found in towns and villages.

**Long-eared Owl** The medium-sized (36cm) long-eared owl can be readily distinguished from the tawny owl by its prominent ear tufts (which are not true ears) and orange-yellow eyes. It is essentially a woodland bird found in mainly coniferous forests, but may also be seen on heaths and marshes, and occasionally in large gardens. The long-eared owl is rarely seen in daylight and is more often heard than seen. Its characteristic call is a long, low 'oo-oo-oo'.

# Raptors

**Falcon** A habit of hovering high in the sky over one spot is the best way of identifying this medium-sized (34cm) falcon. The male's grey head, tail and spotted chestnut mantle are also distinctive. Its call is a shrill 'kee-kee-kee'. It is found in open country and also in parks. Sometimes it will use a large nest box for its nest, which it builds in April or early May. It is a year-round resident in Britain.

**Sparrowhawk** The medium-sized (28-38cm) sparrowhawk—the bird most people refer to when they use the work 'hawk'—is a swift, agile bird, that can often be seen dashing low along open ground, pouncing across hedgerows in a surprise attack, or plummeting down in a closed-wing dive. This bird is often detected by the alarm caused in its prey and it is usually found near woodlands, but sometimes in the suburbs on the outskirts of towns. It is a fearless hunter and has been known to snatch small feeding birds from windowsills and bird tables. The female is larger than the male and may take birds up to the size of woodpigeons, though finches and thrushes are its chief prey. Apart from sighting the actual bird, you may find signs of its kill—a pile of feathers (since these birds always pluck their food before eating it)—or regurgitated, undigested food pellets beneath its favourite perch.

# KINGFISHER

**Kingfisher** If you live near a river or stream, or are lucky enough to have one running through your garden, you may be visited by the colourful kingfisher. Bright blue-green upper parts and contrasting chestnut under parts make this small (16.5cm) bird one of Britain's most unmistakable birds. Kingfishers are swift fliers, and may be seen as a sudden, bright blue flash. They are found near all kinds of fresh water in lowland areas but especially love fast-flowing streams and rivers.

# WAGTAILS

**Grey Wagtail** Yellow under tail coverts and breast, blue-grey upper parts and long tail identify this small (18cm) bird. The male has a black throat in spring. It is often seen near lowland streams, and by faster-flowing hill streams in the breeding season, but it will also travel away from water to feed. In common with other wagtails, it repeatedly wags its tail up and down. Its call is a high-pitched, metallic 'tzitzi', often shortened to an abrupt 'tit'.

**Pied Wagtail** This small (18cm) bird is the only long-tailed, black-and-white bird in the region. Its usual note is a shrill 'tschizzik' and its song, though rarely heard, is a warbling twitter. It is found in open country, and often in association with human habitations, preferably where there is water.

# HERON

Herons are large wading birds with long necks and legs. They feed in shallow water along the coast or on inland waters or marshes. The long neck is held retracted with the legs stretched out behind while the bird is in flight.

**Grey Heron** The large size (90cm), long legs and snake-like neck of the grey heron make it unmistakable. It is often seen standing still in the water, lying in wait for fish, frogs and other small animals, which it stabs with its long beak before swallowing them whole. Its most usual call is a harsh 'kraank' but it uses others, particularly at the nest, which it builds in trees or reed beds. Herons will often visit garden ponds in country areas, searching for food, particularly if goldfish are on the menu! Look for them in the early morning, standing motionless at the pond side, ready to strike.

# Duck

**Mallard** The large (58cm) mallard is the duck that you are most likely to see. It is found on still or flowing fresh-water in parks, large gardens and on canals as well as in wilder locations. The breeding drake's green head distinguishes it from other waterfowl while the duck has a large blue speculum. Only the duck utters the characteristic 'quack'. The mallard is a year round resident in Britain and can become very tame if fed regularly.

# Gulls

Gulls are long-winged seabirds. Adults usually have white under parts and are grey to black on top. Young birds start off brown and gradually develop adult plumage. They nest in noisy colonies and are a familiar sight on buildings and in gardens near the coast.

**Black-headed Gull** The smallest European gull (35-38cm), which is distinguished in summer by its chocolate-brown hood. In winter the hood is reduced to a spot behind the eye. It is widespread inland and may be found nesting around lakes, marshes and even sewage farms. The call is a coarse 'kraahh'.

**Lesser Black-backed Gull** This large (53-56cm) gull is distinguished from the bigger great black-backed gull by its size and more slender build and by its yellow legs in summer. The bird's back is slate-grey rather than black and both sexes look similar. Like young herring gulls, immature birds have mottled brown plumage. The lesser black-backed gull is a bold, aggressive scavenger that eats almost anything, from refuse to other birds' chicks. Its call is similar to that of the herring gull, but harsher. In Britain it is usually a resident bird, staying here all year round.

**Great Black-backed Gull** A powerful bird which is the largest (64-79cm) and rarest of Britain's gulls. More sturdily built than the lesser black-backed gull, with a heavy bill, it is a voracious feeder, frequently attacking smaller seabirds and sometimes even lambs. Whitish-pink legs and dark upper parts also distinguish it from the lesser black-backed.

**Herring Gull** The 'kee-yow kee-yow kee-yow' call of the large (55-66cm) herring gull is one of the most familiar sounds of the British coast. The commonest of the large gulls, the herring gull is distinguished from the similar lesser black-backed gull by its light grey upper parts and pink legs. Its feeding habits range from catching fish to scavenging for food from rubbish tips. It is a year-round resident.

**Common Gull** Despite its name, in Britain the medium-sized (41cm) common gull is only really common in Scotland and north-west Ireland. It can be distinguished from other gulls by its greenish-yellow legs and bill and also by its generally smaller build. The call is a high pitched 'kee-yaa'. It is frequently seen inland and will breed by freshwater as well as on coasts and marine islands. In some areas it is often seen in urban environments, as likely to be seen on town playing fields as they are to be found on a beach.

# LESS COMMON BIRDS

Most of the birds illustrated on the previous pages will be common to gardens throughout the country–you are likely to be visited by blackbirds, sparrows and such like wherever you live. But other less common visitors to the garden will depend very much on the habitats that surround your home. If you live in the centre of farmland you may be visited by birds such as skylarks or pheasants. It's also true that some birds are more common in certain parts of the country than others–you are more likely to see buzzards in northern England than on the south coast for example. In addition to these species, the following pages contain a selection of birds most likely to be seen in and around the urban environment, in parks, towns and cities.

**Skylark** The sustained warbling song of the small (18cm) skylark, given while ascending high into the air and while hovering, is the best way of identifying this bird. On the ground it looks like a nondescript brown bird, with a longish tail and short crest. It is a bird of open grasslands, heaths and farmland and migrates in October or November to southern Europe, sometimes returning as early as the end of February.

**Whinchat** The tiny (12.5cm) whinchat has a prominent white eye stripe. It has a brief warbling song and usually dwells in hilly country where there are isolated trees and bushes. In spring it can often be seen on high perches such as telephone wires. Whinchats are summer visitors to Britain, which appear in April or early May and fly south around the beginning of September.

**Wood Warbler** A yellow breast, white under parts and distinctive yellow eye stripe identify this tiny (12.5cm) bird. It has two songs, a long accelerating trill and a melancholy, repeated 'puu' sound. A summer visitor to Britain, it is usually found in forests and woods, often of beech and oak trees. It spends the winter in Africa, returning at the end of April or the beginning of May. Unusually, the female builds a round covered nest in a tussock on the ground. This bird is larger than the similar willow warbler and is more common in Scotland than elsewhere in the British Isles.

**Marsh Tit** Best distinguished by its voice, the little (11.5cm) marsh tit has a distinctive 'pitchuu' call. Its song is a liquid, bubbling series of repeated notes. It is generally found in broad-leaved woodland and scrub and, though not a common visitor to gardens, is most likely to occur in large gardens or parks with mature trees, particularly in winter. Despite its name, it has no special liking for marshes. It is a resident bird which usually makes its nest in hollows in tree stumps or rotting branches, but it will use nest boxes if nothing else is available. The nest of moss is thickly lined with hairs and may hold anything up to a dozen eggs. These birds remain in breeding pairs throughout the winter and are sometimes seen with flocks of other tits, although they do not usually stray far from their own territory.

**Crossbill** The small (16.5cm) crossbill is unique among European finches for its crossed bill mandibles, which, used to extract seeds from pines, spruces and other conifers, give it its name. The adult male is brick-red in colour, the female a yellowish-green. The chief call is a loud 'chup-chup'. The song is similar to that of the greenfinch with the call intermingled. Crossbills are mainly found in coniferous forests, especially of spruce and fir, often in parties. They are sometimes seen in towns and villages. The young are born with straight bills which become crossed only after they have left the nest, roaming the countryside in the company of their parents in search of food.

**Siskin** The siskin is one of the smallest members (12cm) of the finch family. Predominantly yellow-green, the male is distinguished by his black crown and chin and yellow rump. The siskin's main call is a clear 'tsuu' or 'tsyzing' given on the wing. It is a winter visitor to most of Britain, but is resident in Ireland and Scotland. It is most often found in mixed or coniferous woods which it leaves in winter to form flocks, roaming hedgerows, birch and alder groves in search of seeds.

**Linnet** A tiny (13.5cm) finch, the male linnet is distinguished by its chestnut mantle and, in summer, by its crimson crown and breast. The female is less strongly marked and does not acquire the crimson colouring. Its song is a varied twittering, the usual flight call being a rapid 'chichichichit'. It is found in areas with plenty of low trees and bushes and on more open land in winter.

**Buzzard** The large (51-56cm) buzzard is a raptor that shows great variation in its colouring, which ranges from pale to very dark, but is usually dark brown above and pale brown below. Its high-pitched, gull-like cry can be heard as it soars high above wooded hillsides and forests, scanning the ground for its prey.

Buzzards are resident birds in Britain, more commonly seen in northern and western Britain and in northern Ireland. They were heavily persecuted in the last century, but since then their numbers have increased. They are most often seen circling over woodland or sitting motionless on a post at the side of the road. They will often wait on the ground outside the hole of a mouse or other small mammal, sometimes remaining there for hours in order to catch their prey. Rabbits are another important source of food. Buzzards are not normally seen in towns, but occasionally a juvenile bird may stray off course during migration.

**Pheasant** The world's most widespread gamebird, the large (male 66-89cm, female 53-63cm) pheasant is found in fields, heathland and forests, and also large parks. It was probably introduced to Britain in Roman times. The long tail is the best distinguishing feature of both male and female birds. The male is a rich copper colour, with a metallic-green head and red wattles around the eyes. The female's plumage is a light brown colour. When startled, a pheasant can shoot upwards at great speed, but will often prefer to run for cover rather than take to the air. The voice is a far-carrying 'kok-kok-kok'. These attractive birds are a common sight on roadsides and in fields, particularly during the autumn game season.

Swans are the largest of Britain's waterfowl. Majestic white birds, which swim gracefully on rivers and lakes, they are waddling and ungainly on land and require a long take-off run to take flight on their powerful wings.

**Mute Swan** The huge (152cm) mute swan is a year-round resident in Britain. The adult has an orange bill and, at rest, the neck is held in the curved 'swan-neck' posture. In flight, the wings make a throbbing sound. Adults hiss threateningly if annoyed. The nest is a mound of reeds and other plants built near water. They are a common sight wherever there is water, even in the middle of towns, frequenting park lakes, rivers and smaller ponds.

Geese are large birds, capable of flying powerfully for long distances. They generally have longer necks in relation to their bodies than do most ducks. They are usually seen in noisy flocks and feed mostly on land, although they are also good swimmers.

**Canada Goose** Any black-necked goose spotted on ponds, lakes or marshes is almost certain to be the large (92-102cm) canada goose. Originally a species which was introduced from North America, it is now well established in Britain. The nest, made from vegetation near water, will be defended aggressively. The call is a loud, trumpeting honk 'grunk-grunk' sound. In some places this goose has become so numerous that it is considered a pest. It is a familiar sight in town parks and can often be seen–and heard–flying in noisy flocks over the urban environment.

**Coot** No other black waterfowl has the white forehead that distinguishes this medium-sized (38cm) rail. The coot's head and neck are glossier than the rest of the body and the lobed feet, unlike those of other rails, can be seen when the bird feeds on land. The coot flies reluctantly, gaining the air with difficulty from the water. It can stay submerged for up to half a minute when feeding. Males are territorially aggressive, often chasing each other off as well as any other intruder, before taking cover amongst the waterside vegetation. The call of the coot is loud and distinctive, and sounds like 'kook' or 'kewk'. Coots are common throughout most parts of the British Isles where they will be found on large bodies of fresh water, including park ponds and lakes and even concrete reservoirs. They are often confused with the equally common moorhen (see opposite) but lack the moorhen's bright red bill and forehead.

**Moorhen** The moorhen is a medium-sized (33cm) rail whose main distinguishing features are a white band along its flank and its red bill and forehead. Out of the water, the green legs may be seen with a characteristic red 'garter' at the base. The moorhen is commonly found around watersides with plenty of cover along the edges. It is often seen in town parks where there is water as well as along the edges of reservoirs and muddy ditches. If disturbed, the moorhen may dive, leaving only its bill visible above the surface of the water.

Moorhen

**Great Crested Grebe** In the breeding season, the large (48cm) great crested grebe adults have conspicuous dark double-crests and a chestnut and black neck ruff. These are used to full effect in an elaborate courtship display, involving much head-shaking and fluffing out of feathers. The nest is made of waterweed anchored to reeds. The adults carry the young chicks on their backs. These birds were quite infrequent before man-made lakes provided them with an ideal habitat.

**Oystercatcher** The large (43cm) oystercatcher is a distinctive black-and-white wader, distinguished from other large shorebirds by its long red bill and pink legs. In winter it has a white half-collar on the throat and, in flight, its white wing bar can be clearly seen. Flocks can be seen along sandy and rocky beaches and estuaries, where food is plentiful. In autumn and winter, flocks of hundreds of birds are common on good sites. In Scotland they are also seen on the shores of lochs or feeding on arable land. In recent years these primarily coastal birds have taken to roosting on rooftops in seaside towns and searching for invertebrates on nearby playing fields and other open spaces. They also breed on relay structures set up for the North Sea oil industry. The call is a high-pitched 'kleeep-kleeep'.

**Tufted Duck** The tufted duck is a medium-sized (43cm) bird. The drake's drooping black crest is unique to the tufted duck; the duck's crest is shorter and brown. Since its introduction in 1849, the tufted duck has become Britain's commonest species of diving duck. The drake occasionally utters a low whistle, the duck makes a growling noise. It is a common species on man-made lakes and ponds.

**Pochard** The pochard is a medium-sized (46cm) duck. The drake, easily identified by a chestnut head and neck and grey upper parts, is the only duck in this region with such colouring. A freshwater diving duck, pochards are rarely seen on salt water. The nest is built on ground near water or on a platform of water plants.

# Encouraging Garden Birds

On the following pages you can see some of the simplest ways to make your garden bird-friendly. There are other ways of attracting birds, however. If you have a male robin in the neighbourhood, for instance, you will often be able to witness his fierce, territorial nature if you sit a little model bird out, as long as the model has a red breast—his stimulus to attack!

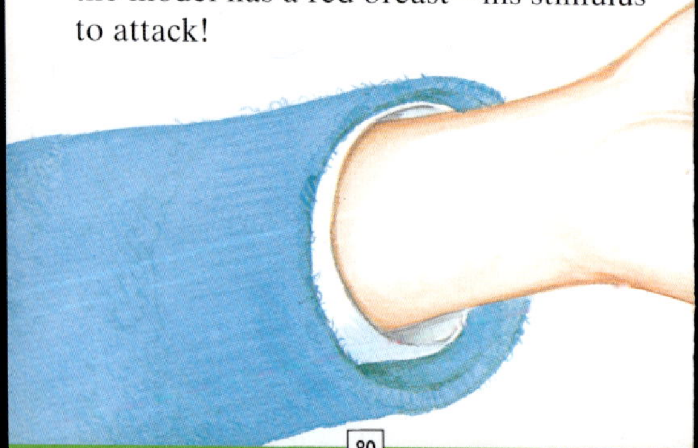

***Some birds, such as house sparrows, can become tame enough to take food from your hand if fed regularly!***

# Bird Tables & Feeding

Food becomes scarce for birds during the autumn and winter and it can be a real help to them if you provide them with additional nourishment. Of course, this also has the added advantage of encouraging them to visit your garden more regularly!

While birds such as wrens seem to enjoy feeding on the ground, most birds will prefer the comfort and safety of a bird table such as the one shown opposite. One problem birds often face in gardens is cats, and this can be helped by putting a baffle around the pole on which the table is placed, or possibly even spraying cat repellant around the base. A small bowl of water on the table will usually be appreciated, too!

Baffle

In putting out food for wild birds, the best thing to do is to provide as wide a range of food as possible. This will encourage the widest range of birds into your garden. Kitchen scraps such as cheese, bacon rind or even the bone of the weekend joint are often popular, as are dried fruit, porridge oats, nuts, apples, cereals and half coconuts hung upside down. You can also buy wild bird food which contains a wide range of seeds to encourage the seed eaters. This food can be either placed in hanging bird feeders or on the bird table. One recipe birds seem particularly fond of is 'bird pudding': melt some dripping and mix in seeds, cake crumbs, cheese, oats, unsalted peanuts or whatever is available and pour into yogurt pot moulds. When cool, tip out the pudding and place it on the table. Delicious!

## NEST BOXES IN THE GARDEN

An excellent way of encouraging birds is to buy or build a nest box. As with bird tables the main rule is simple: if you put one up, keep it well out of the way of predators such as cats! One type of nest box is shown overleaf. Another approach might also prevent damage being done by certain birds such as house sparrows who are fond of nesting in places such as gutters. They will happily nest instead in a simple 'tunnel box' made out of three straight pieces of wood fastened under the eaves. Of course, it is not strictly necessary to buy or build an actual box—birds will nest in many unusual places, but if you wedge something like an old saucepan in a tree, always make sure it is well out of the prevailing wind!

Nails and screws | Leather hinge | Side 1 | Side 2

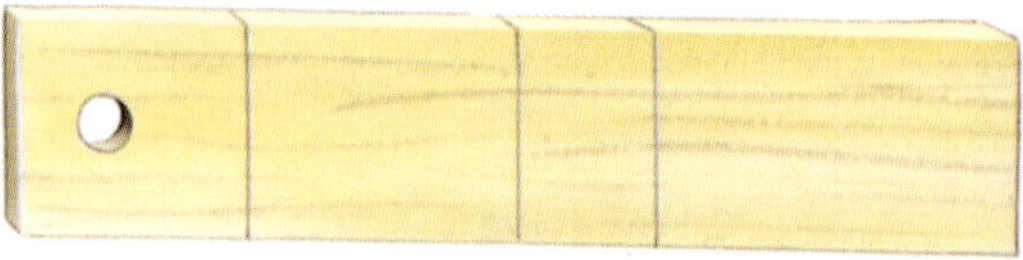

Front | Lid | Base | Back

You can make a nest box from a single plank of wood, about 15cm wide and 18mm thick. Cut it up as shown and drill a large hole in the front– a 30mm diameter will be suitable for tits and other small birds. Nail the pieces together as shown and use a strip of leather or rubber tacked to the lid to make a hinge. Fix the box to the trunk of a tree, well off the ground and out of the glare of the sun using two strong screws.

## Garden Design & Plants

If you enjoy the idea of making your garden into a haven for as wide a range of birds as possible, there are a number of things you can do. First, it is a useful idea to create as wide a range of habitats as possible—the ideal garden design might include an area of tangled undergrowth, as well as a variety of trees, shrubs and more open areas. Building a small pond will also encourage a number of birds to visit although, unless you have quite a sizeable pond, you are unlikely to get many ducks, geese or swans living there, although you may get the occasional visitor passing through. Fish in the pond may attract birds such as herons. Secondly, it is important to consider which particular plants you will grow.

Plants that provide food for birds are ideal. Sunflowers provide a good meal for birds such as nuthatches and finches, while fruit-eating birds may be grateful for berry-bearing shrubs and trees such as cotoneaster, hawthorn, holly and privet. Goldfinches seem to love pulling the seeds out of thistle heads, although thistles may not be your idea of the perfect plant for your garden!

# GLOSSARY

**brood:** young birds that have all hatched out from the same clutch of eggs.

**clutch:** eggs in a nest that have all been laid by the same bird.

**dabbling ducks:** ducks that feed on or near the surface of the water.

**display:** signals sent by a bird, usually male, either to attract a mate, or to warn off other males.

**down:** soft feathers covering a newly-hatched bird.

**field mark:** a feature of a bird's plumage that aids its identification, for example, the coloured speculums of ducks.

**introduction:** a bird that has been brought in from outside its normal range and released into the wild.

**juvenile:** a young bird that has taken on its first covering of true feathers, as distinct from the down that covered it while it was a nestling, but has not yet acquired its adult plumage.

**migration:** the movement of a population of birds or other animals from one area to another at around the same time each year followed by a later return to the original area. Birds migrate, often over long distances, to ensure a regular supply of food and suitable breeding conditions.

**moult:** the natural loss and replacement of feathers as they wear out.

**nocturnal:** active at night, most owls are nocturnal, for example.

**plumage:** the name given to the bird's entire covering of feathers.

**raptor:** a day-flying bird of prey, such as a sparrowhawk or kestrel.

**resident:** a bird that stays in the same area throughout the year.

**roost:** a group of birds gathered together in a communal sleeping spot. Also means to rest or to sleep.

**soaring:** circling flight using uprising warm air currents, called thermals, to stay aloft rather than wingbeats. Eagles and buzzards are typical examples of soaring birds.

**species:** an animal or plant of a particular kind that can be distinguished from all other living things. Opposite sex members of a species can breed together to produce a new generation that will be similar in appearance to their parents.

**speculum:** the coloured patch on a duck's wing.

**territory:** an area that is defended by a bird or pair of birds from others of the same species. The territory is usually, but not always, centred on the nest site.

**wing bar:** a bar across the wing that differs in colour from the surrounding feathers; often a good guide to identification.

# INDEX


Blackbird 12
Blackcap 21
Bullfinch 43
Buzzard 72

Carrion Crow 36
Chaffinch 42
Chiffchaff 22
Cuckoo 49
Coot 76
Crossbill 70

Duck, Tufted 79
Dunnock 39

Firecrest 25
Flycatcher, Spotted 26

Goldcrest 24
Goldfinch 45
Goose, Canada 75
Grebe, Great Crested 77
Greenfinch 44
Grey Heron 60
Gull, Black-headed 62
Common 65
Great Black-backed 64
Lesser Black-backed 63

Hawfinch 46
Hoopoe 27

Jackdaw 36
Jay 34

Kingfisher 58

Linnet 71

Magpie 35
Mallard 61
Martin, House 52
Moorhen 77

Nightingale 17
Nuthatch 40

Owl, Barn 54
Little 54
Long-eared 55
Tawny 55
Oystercatcher 78

Peregrine Falcon 56
Pheasant 73
Pochard 79

Redpoll 46
Redstart 16
Redwing 14
Robin 15
Rook 37

Siskin 71
Skylark 67
Sparrow,
  House 38
Sparrowhawk 57
Starling 18
Swallow 53
Swan, Mute 74
Swift 52

Thrush, Mistle 13
  Song 13
Tit, Blue 30
  Coal 28
  Great 31
  Long-tailed 29
  Marsh 69
Treecreeper 41

Wagtail, Grey 59
  Pied 59
Warbler,
  Garden 19
  Willow 23
  Wood 68
Waxwing 32
Whinchat 67
Whitethroat 20
Woodpecker,
  Great Spotted 51
  Green 50
  Lesser Spotted 51
Woodpigeon 48
Wren 33

Yellowhammer 47